Teachers Notes for "Money Activity Book"

"Money Activity Book" was created with the following objectives in mind.

- to develop the ability to distinguish between types of coins
- to recognize the value of each individual coin
- to develop the ability to determine the worth of a collection of coins
- to provide an understanding of coins and their uses
- to utilize problem solving skills when dealing with the interpretation of coins

It is the author's hope that, through these objectives, the "Money Activity Book" will meet the need for supplemental material that teaches the concept of money. The book was designed so that elementary teachers can use the materials in conjunction with concepts introduced in their curriculum. It introduces children to coins and encourages them to become familiar with them. The material then progresses through operations and other means by which money is used everyday.

The decision to limit the money concepts in the book to coins was made with the awareness that, once the ideas of cost and making change were understood, they could be expanded at a later date to deal with larger amounts of money. It was felt that children first need to be made comfortable with coins.

"Money Activity Book" can be used in a variety of ways in the classroom. Many of the exercises can be utilized as extra prac- tice, while others lend themselves to use as enrichment. Exercises do not need to be used with the entire class, although you may wish to do so to illustrate your lesson. When working with small groups, a teacher may want to do the exercises orally, perhaps simulating the activity a cashier performs when counting back change. Similarly, exercises could be employed as a warm-up, or wrap-up, to a class period. Another application of the material is reinforcement of skills.

Although the directions on many of the pages indicate that the students should stamp the solution, use of coin stamps is op- tional. Teachers may want to use real coins, thus allowing the lesson to relate more closely to the "real world." Another option which students may enjoy is to have them draw pictures of the coin needed where stamping is indicated. A third choice is to duplicate copies of pages of coins, cut out the pictures, and let the students paste in the answers.

All these ideas are offered as suggestions. As each group of students is different, classroom teachers can decide how best to fit the materials into their curriculum. Which ever way they go, it is the hope of the author that the "Money Activity Book" helps to make their job easier.

⚠ WARNING:
CHOKING HAZARD - Small parts.
Not for children under 3 years.

MONEY ACTIVITIES BOOK

Written by Sherry Wolf

It is important to be able to recognize the coins you use everyday as money. There are two sides to all coins—heads and tails. Use the guide below to identify them.

The most common coins are:

Head   Tail

$.01 OR **1¢**

Head   Tail

$.05 OR **5¢**

Head   Tail

$.10 OR **10¢**

Head   Tail

$.25 OR **25¢**

Not so common but also used:

Head   Tail

$.50 OR **50¢**

Use the guide sheet to answer the following questions about the coins.

1. President Roosevelt is on the head of _____

2. What words appear on the head of every coin _____

3. Thomas Jefferson's home—Monticello—appears on the tail of a _____

4. What are the two number names of a penny _____ or _____

5. Whose picture is on the head of a quarter? _____

6. An eagle is on the tail of both a _____ and a _____

7. What building do we find on the back of a penny _____

 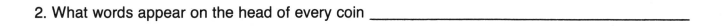  

3

Color the space under each _type_ of coin a different color, i.e. all pennies will be same color, etc.

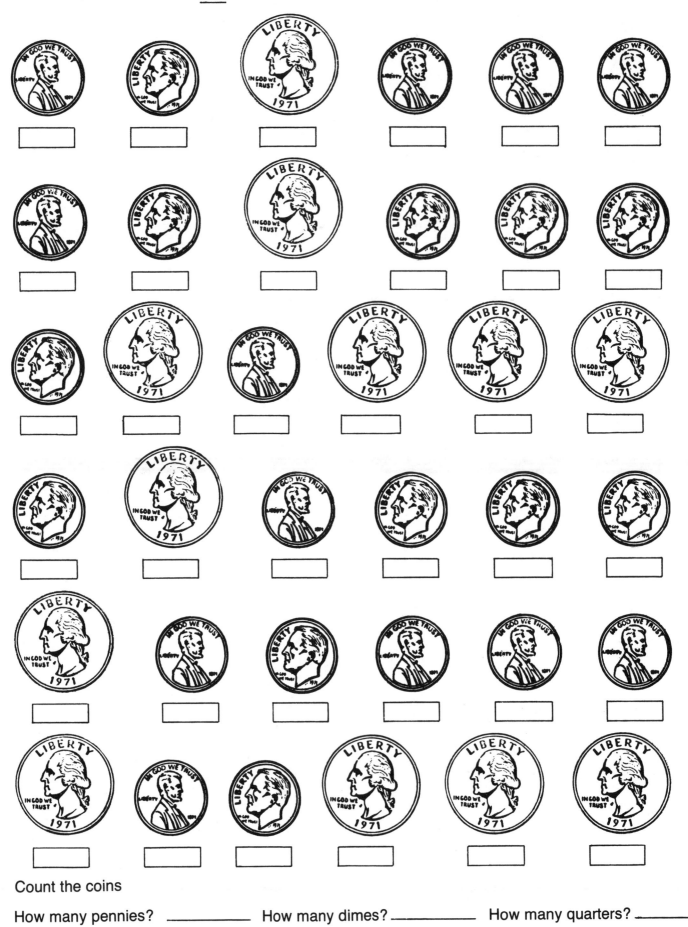

Count the coins

How many pennies? _____ How many dimes? _____ How many quarters? _____

Color the space under each *type* of coin a different color.

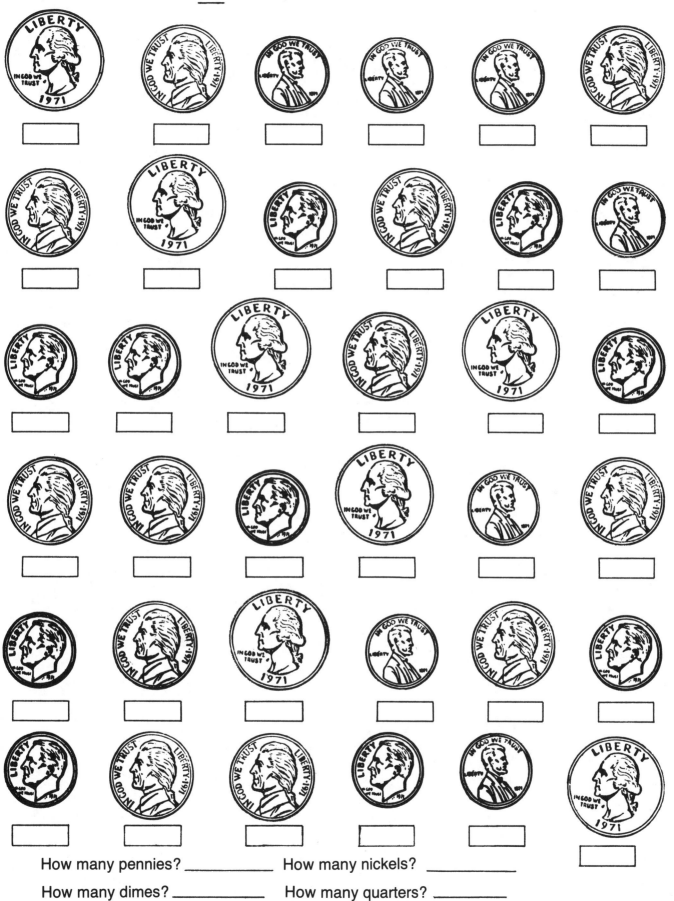

How many pennies? _____ How many nickels? _____

How many dimes? _____ How many quarters? _____

Color the square under each type of coin a different color.

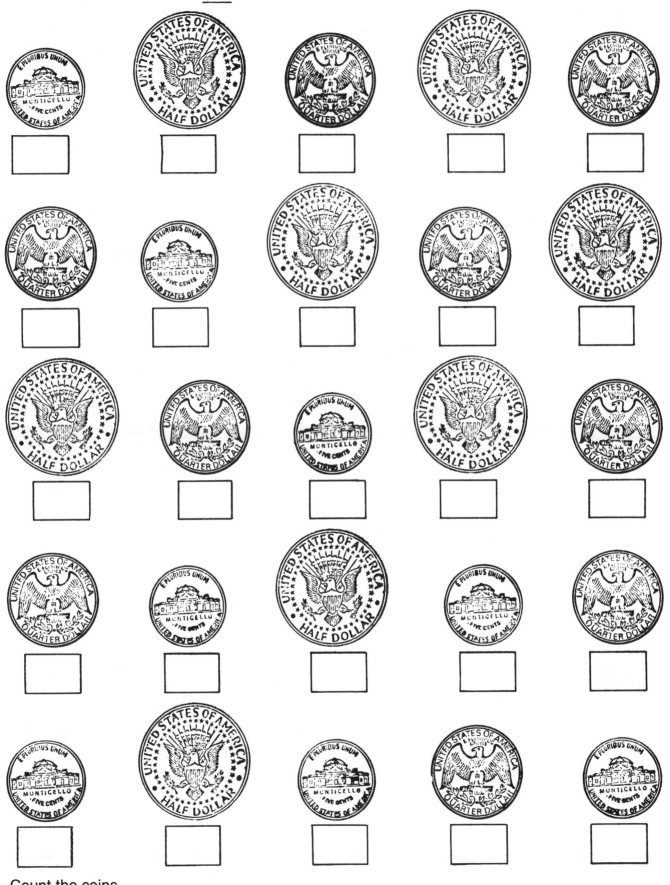

Count the coins.

How many nickels? _____ How many quarters? _____ How many half-dollars? _____

6

Count the value of the coins from left to right. As you count, write the value of the group to that point. The first one is started for you...

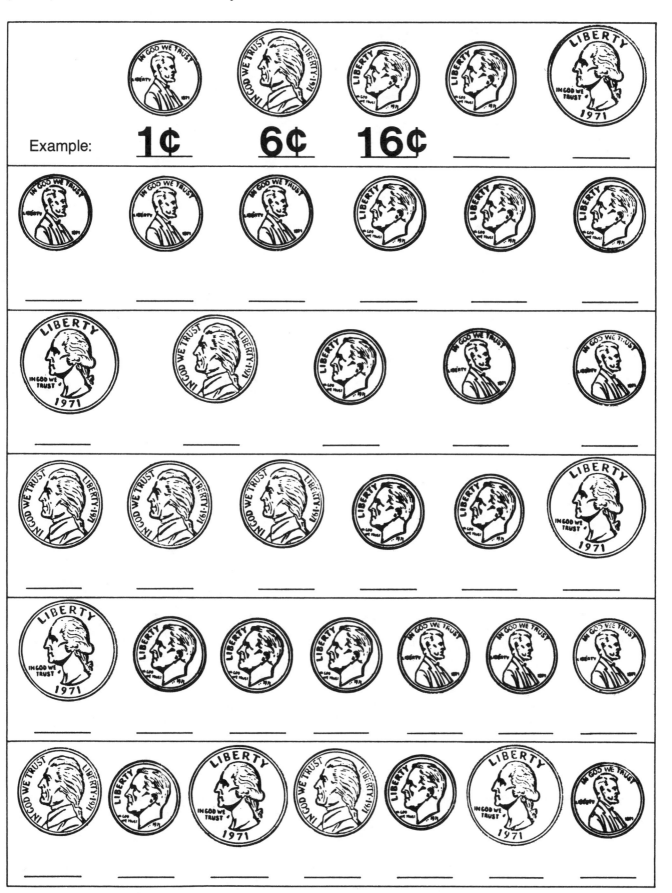

Example: **1¢**   **6¢**   **16¢**   _____

Count the value of the coins from left to right. As you count, write the value of the group to that point. The first one is started for you...

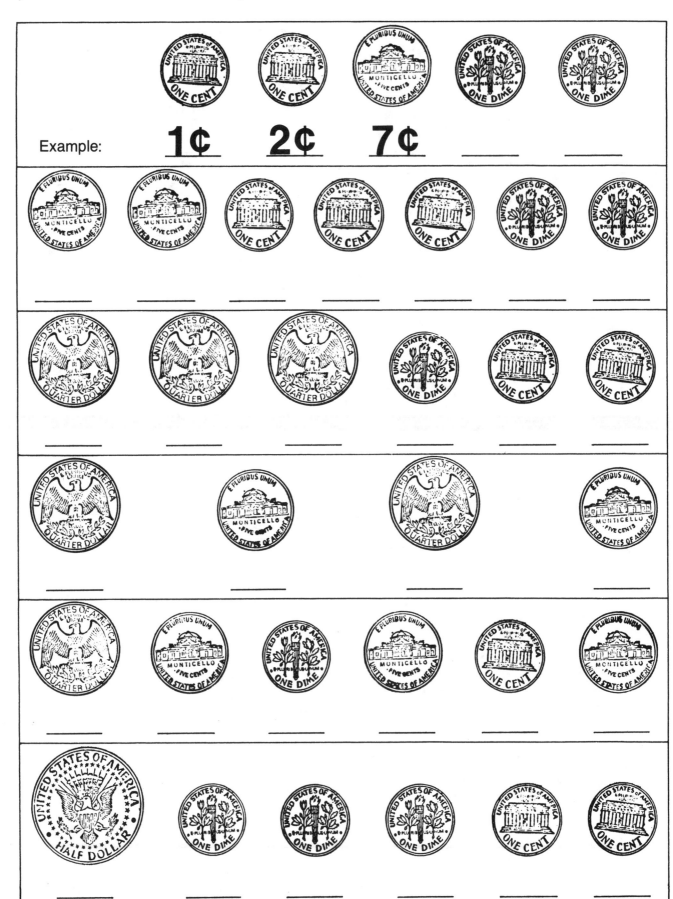

Example: **1¢**   **2¢**   **7¢**   _____   _____

Count the value of the coins from left to right. As you count, write the value of the group to that point. The first one is started for you...

Example:

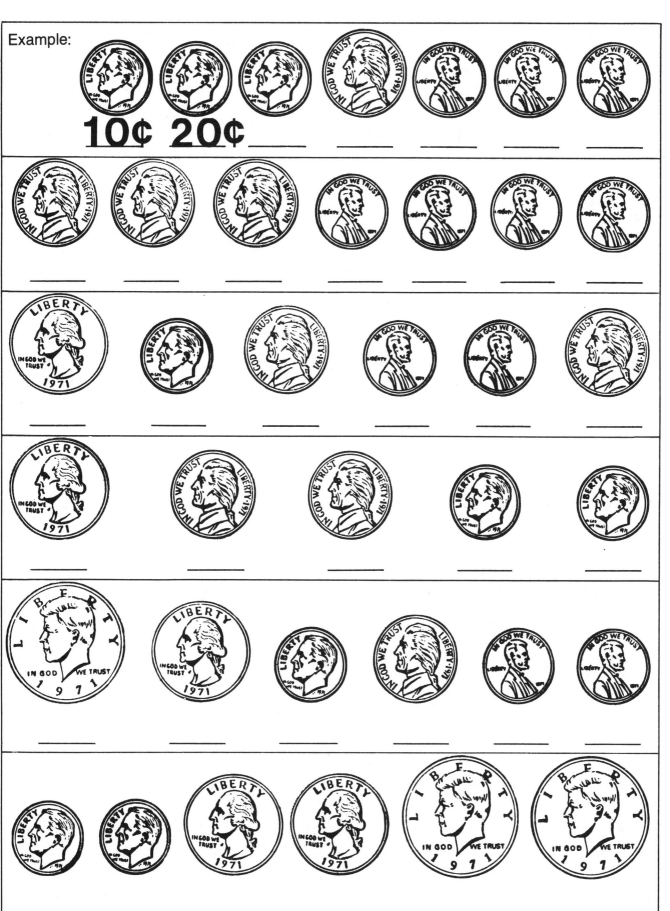

10¢ **20¢** ___ ___ ___ ___ ___

___ ___ ___ ___ ___ ___ ___

___ ___ ___ ___ ___ ___

___ ___ ___ ___ ___

___ ___ ___ ___ ___ ___

___ ___ ___ ___ ___ ___

Count the value of the coins from left to right. As you count, write the value of the group to that point. The first one is started for you . . . .

Example:

5¢   15¢   40¢   ___   ___

___   ___   ___   ___   ___   ___   ___

___   ___   ___   ___   ___   ___

___   ___   ___   ___   ___   ___

___   ___   ___   ___   ___   ___

___   ___   ___   ___   ___

Fill in the blanks.

1. Type of coin? _____

How many? _____

What is the value of the group? _____

2. Type of coin? _____

How many? _____

What is the value of the group? _____

3. Type of coin? _____

How many? _____

What is the value of the group? _____

4. Type of coin? _____

How many? _____

What is the value of the group? _____

Fill in the blanks.

1. Type of coin? _____

How many? _____

What is the value of the group? _____

2. Type of coin? _____

How many? _____

What is the value of the group? _____

3. Type of coin? _____

How many? _____

What is the value of the group? _____

4. Type of coin? _____

How many? _____

What is the value of the group? _____

Stamp the answer with the least amount of coins.

REMEMBER TO WATCH THE OPERATIONS!

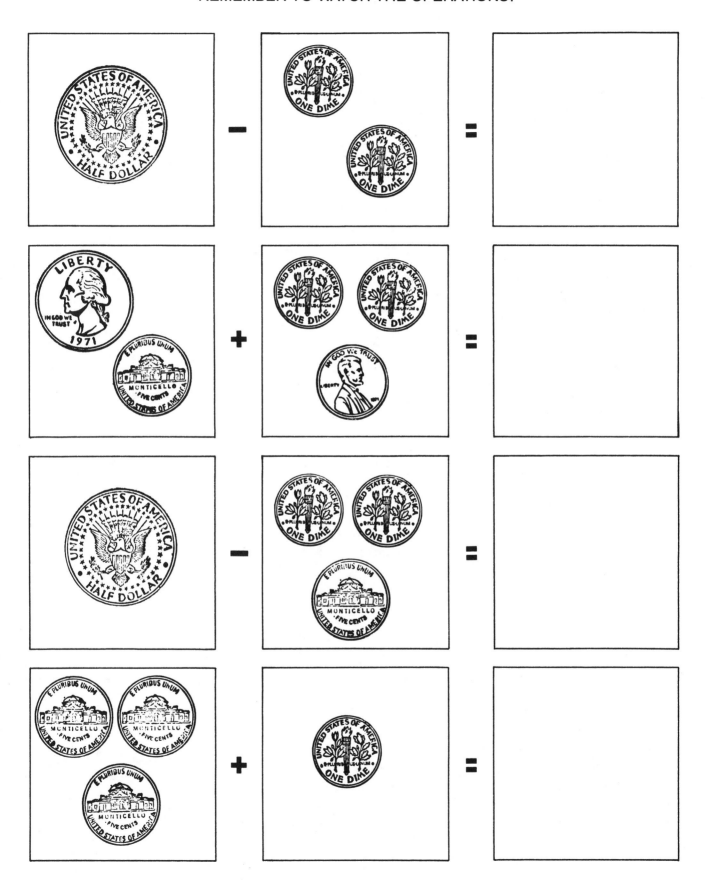

Stamp the answer with the least amount of coins.

REMEMBER TO WATCH THE OPERATIONS!

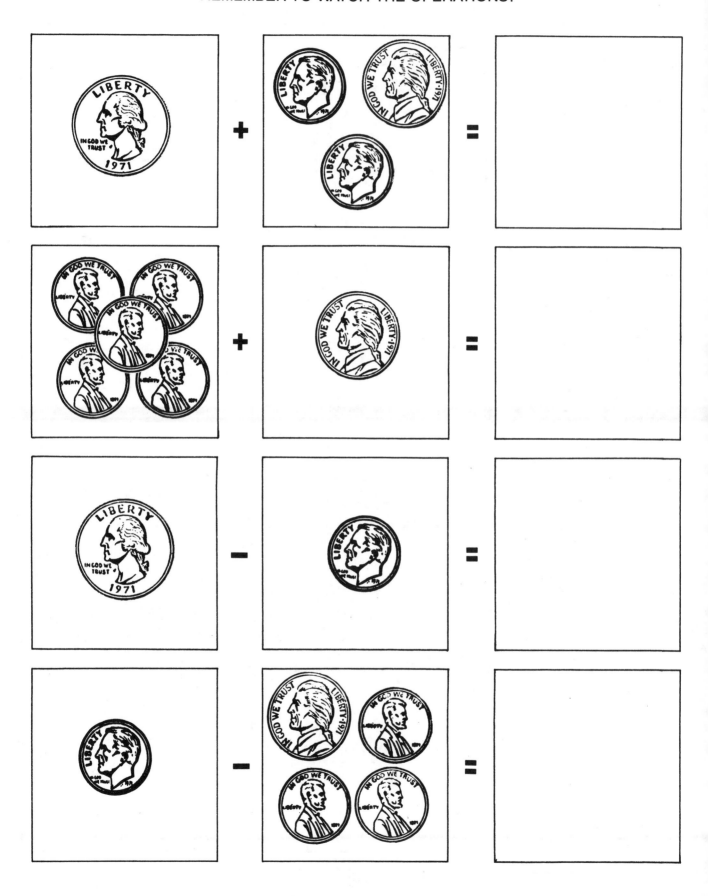

Stamp the answer with the least amount of coins.

REMEMBER TO WATCH THE OPERATIONS!

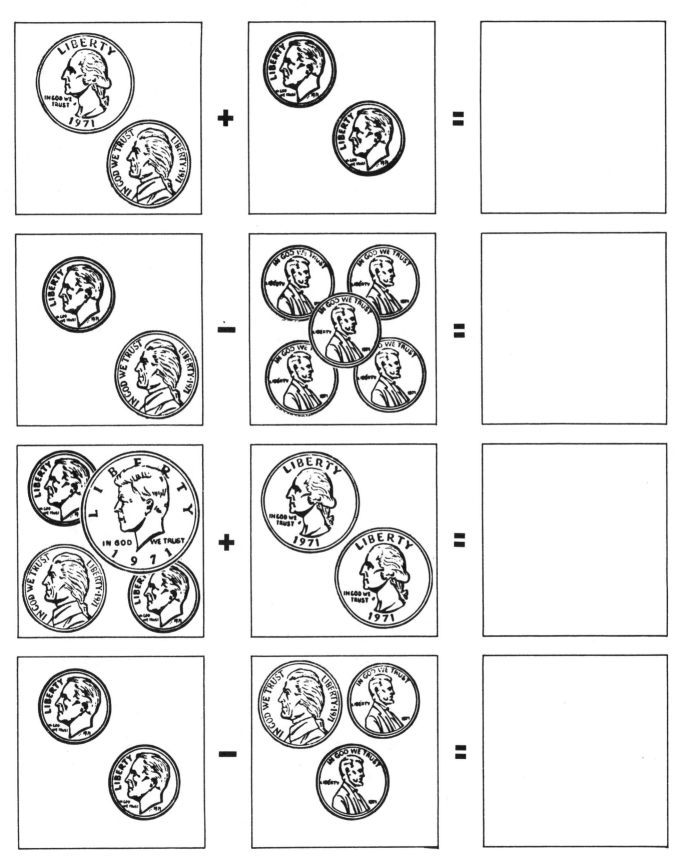

You decide the coin values and whether to add or subtract to reach the pictured result.

		= $0.30
		= $0.15
		= $0.50
		= $0.11

You decide the coin values and whether to add or subtract to reach the pictured result.

		= $0.35
		= $0.06
		= $0.10
		= $0.63

Perform the indicated operation and stamp the answer using the least amount of coins.

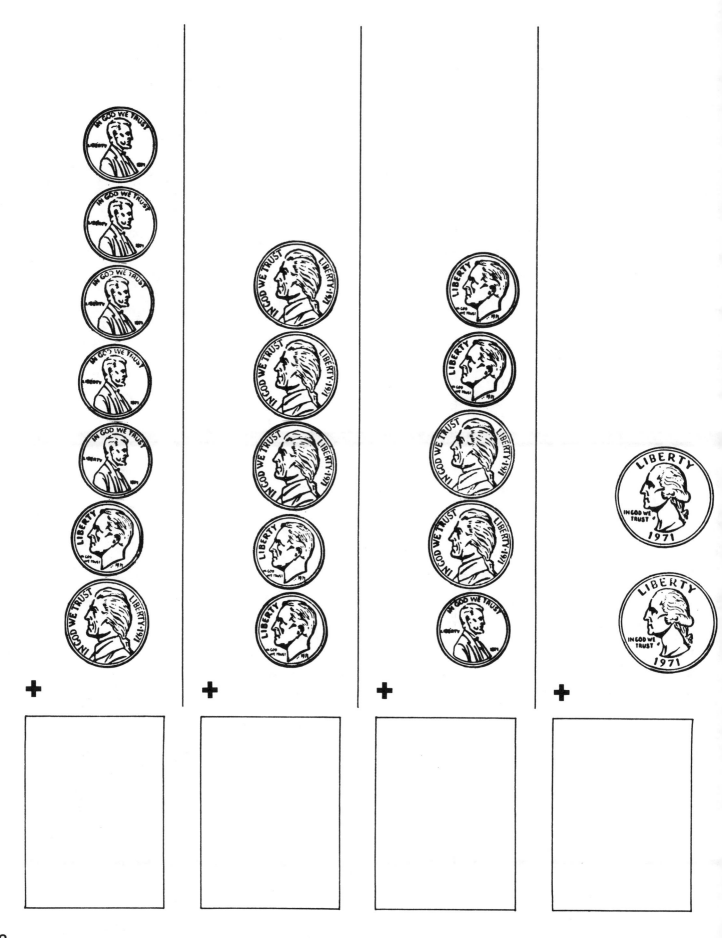

Match the coins to their number name and value by drawing a line to the correct boxes.

SIXTY-FIVE CENTS		$0.53
THIRTY-SIX CENTS		$0.30
THIRTY CENTS		$0.36
FIFTY-THREE CENTS		$0.65

Write the value of each group of coins in the space provided.

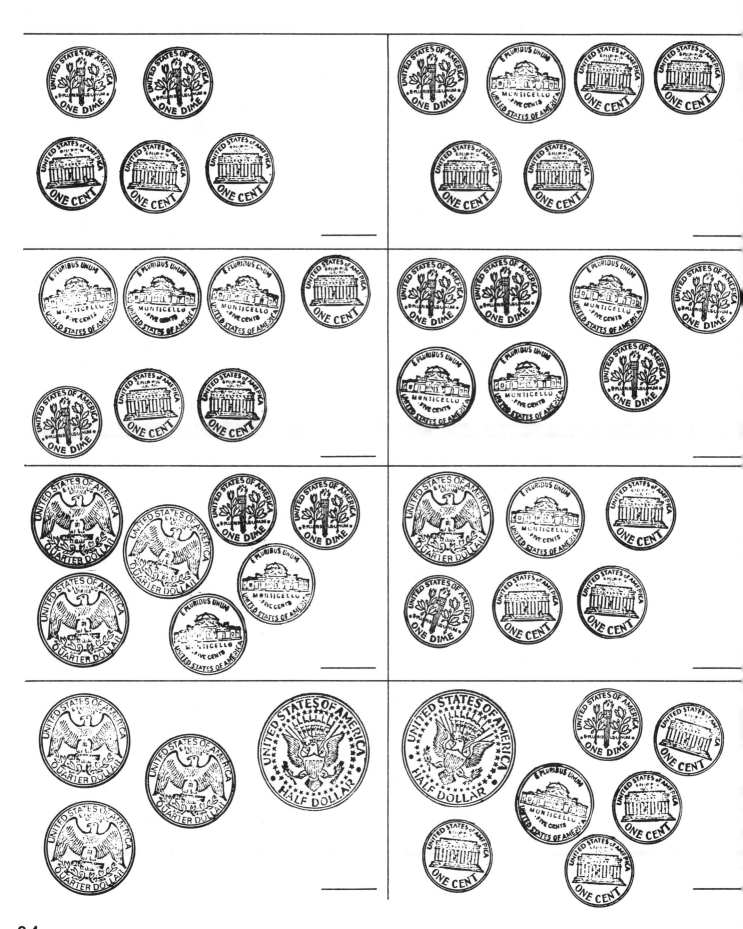

24

Write the value of each group of coins in the space provided.

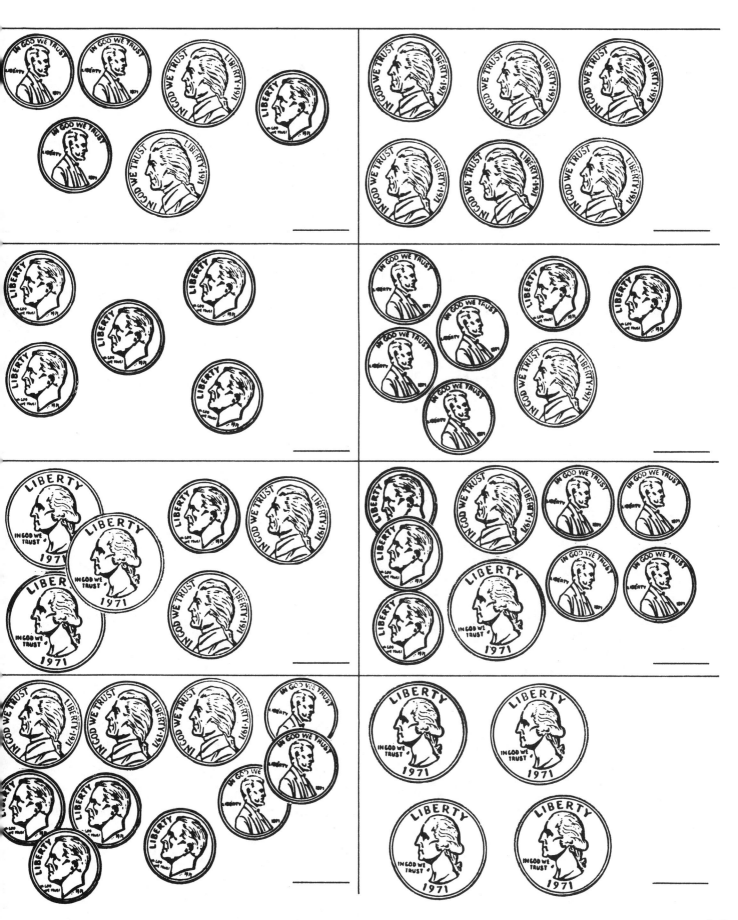

25

Write the value of each group of coins in the space provided.

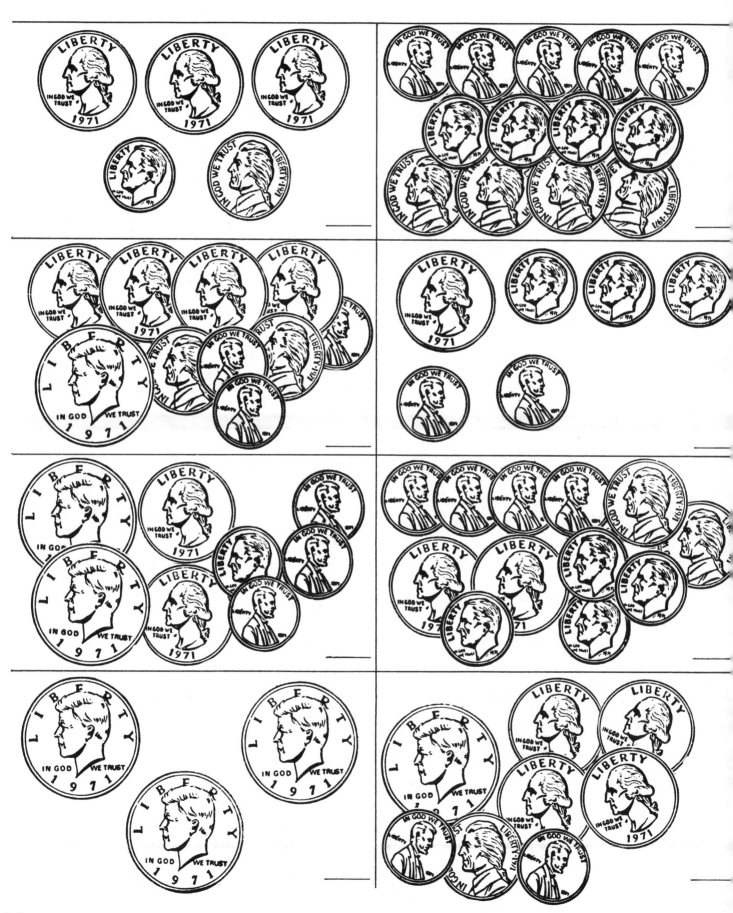

26

Stamp a collection of coins to illustrate the value:

8¢	**16¢**
35¢	**43¢**
55¢	**76¢**
95¢	**84¢**

Stamp a collection of coins to illustrate the value:

33¢	**76¢**
55¢	**63¢**
99¢	**17¢**
26¢	**$1.06**
35¢	**89¢**

Stamp a collection of coins to illustrate the value:

11¢	**36¢**
59¢	**67¢**
48¢	**32¢**
65¢	**$1.25**
98¢	**$1.05**

Place an "X" over the box which has the greater value in each set.

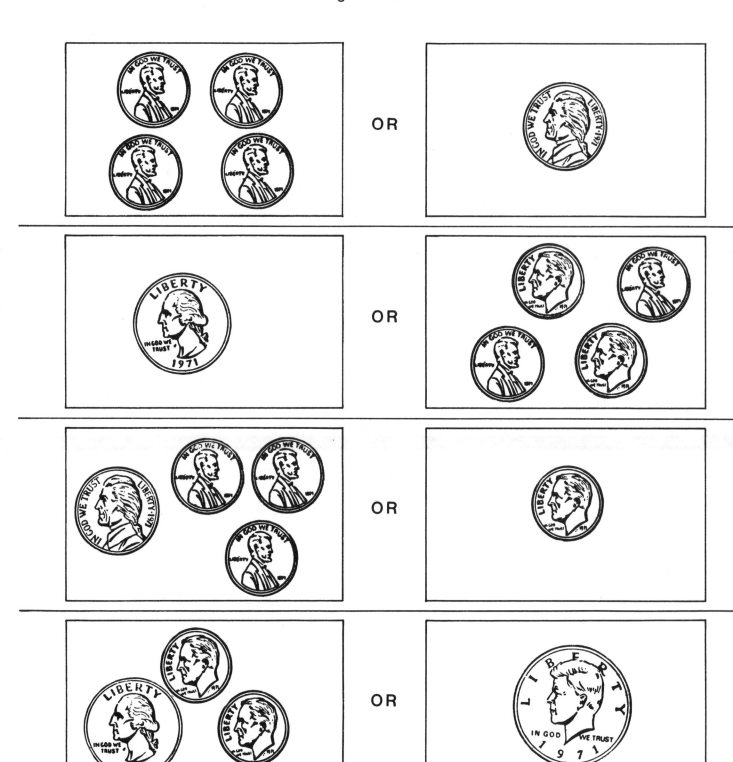

1. Write the value.

2. Answer greater than (>); less than (<); or is equal (=) to make a true sentence.

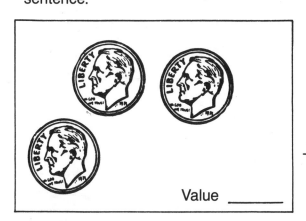

Value _____

Symbol _____

Value _____

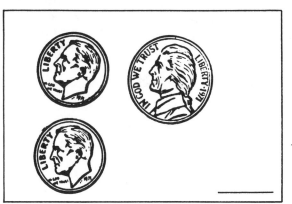

_____

_____

_____

_____

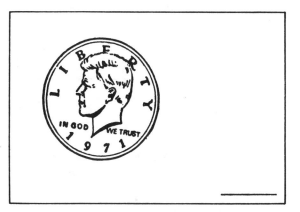

_____

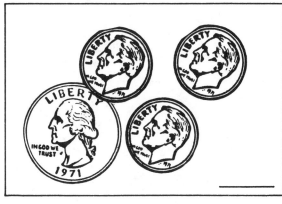

_____

_____

Write the value of each group of coins

then use the symbols , < or = to describe each.

 Value _____   Symbol _____    Value _____

 _____    _____

 _____    _____

 _____    _____

 _____    _____

32

Stamp a collection of coins to make a true statement.

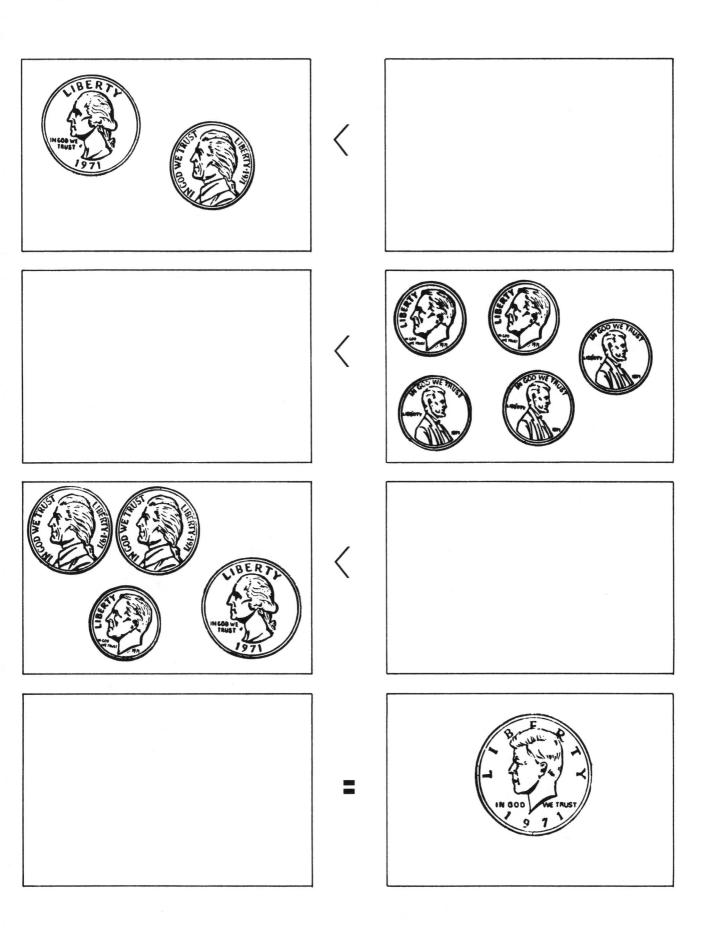

Stamp a collection of coins to make a true statement.

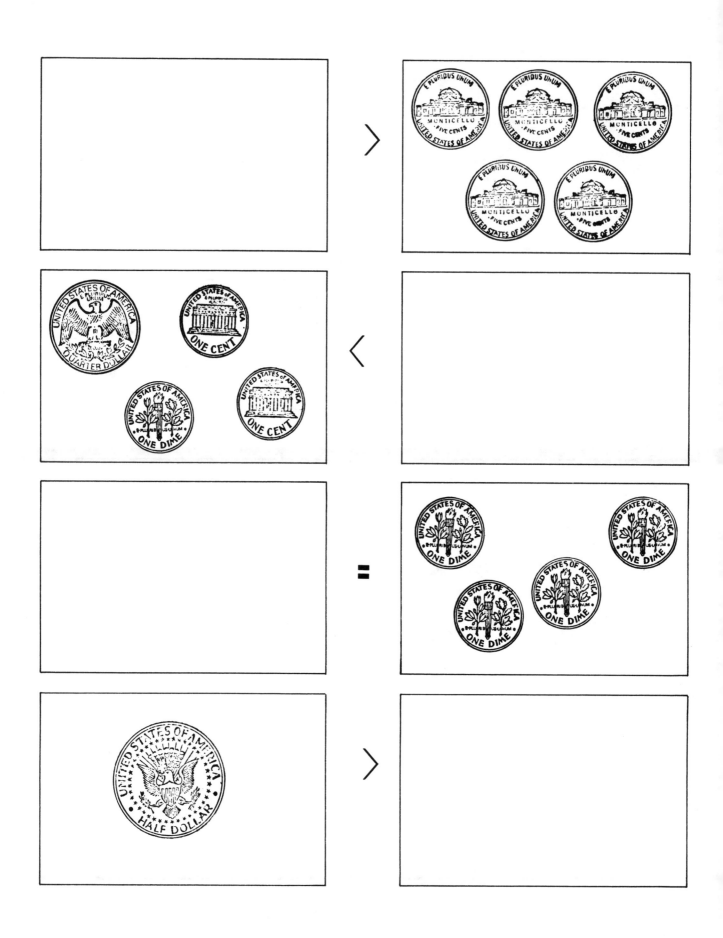

Stamp the result, using a single coin.

 X **5** =

 X **25** =

 X **2** + =

 X **5** =

 X **2** =

 X **10** =

 X **10** =

Stamp two or less coins to equal the products.

 X **6** =

 X **15** =

 X **30** =

 X **3** =

 X **4** =

 X **10** =

Stamp three coins you would receive by multiplying the coin by:

 X **9** =

 X **5** =

 X **15** =

 X **8** =

 X **13** =

 X **6** =

 X **7** =

37

Stamp the answer, using the fewest coins possible.

 X **50** =

 X **12** =

 X **8** =

 X **5** =

 X **2** =

 X **20** =

 X **3** =

Help Fred share his wealth. Divide these amounts into equal groupings of coins.

Nice Guy John always _equally_ _shared_ his good luck with his friends. Help him divide his money by stamping the amount each boy would receive in the box.

1. $0.75 for money a lawn

2. Found $0.20

3. $0.36 for running an errand

Certain values of coins can be _divided_ in many ways.

Divide up the amount below so it fits evenly in each of the situations shown.

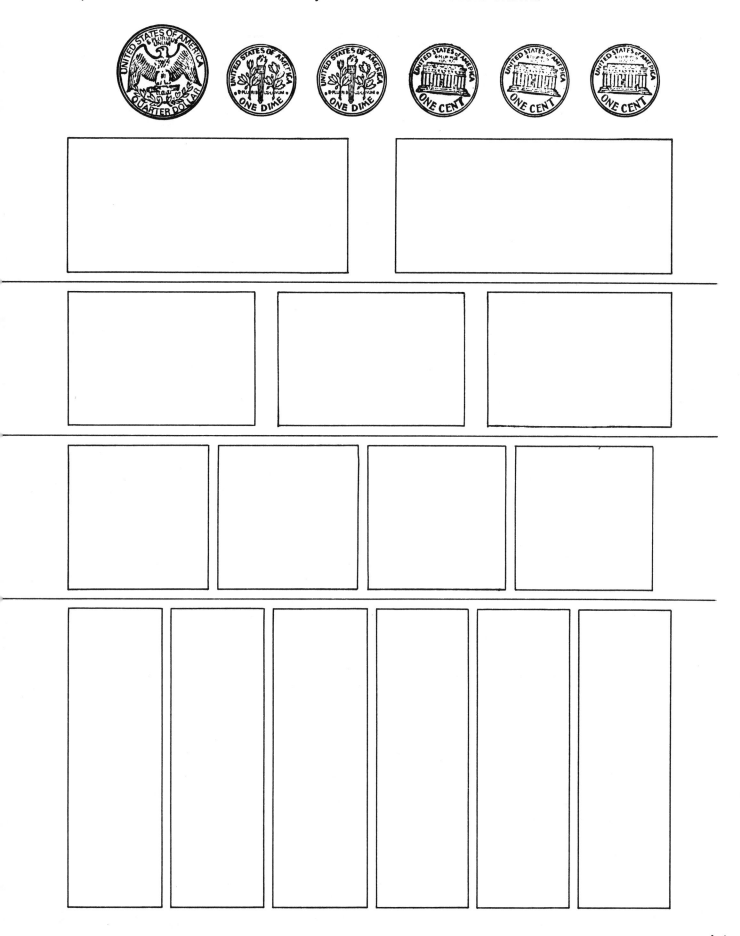

Certain values of coins can be _divided_ in many ways.

Divide up the amount below so it fits evenly in each of the situations shown.

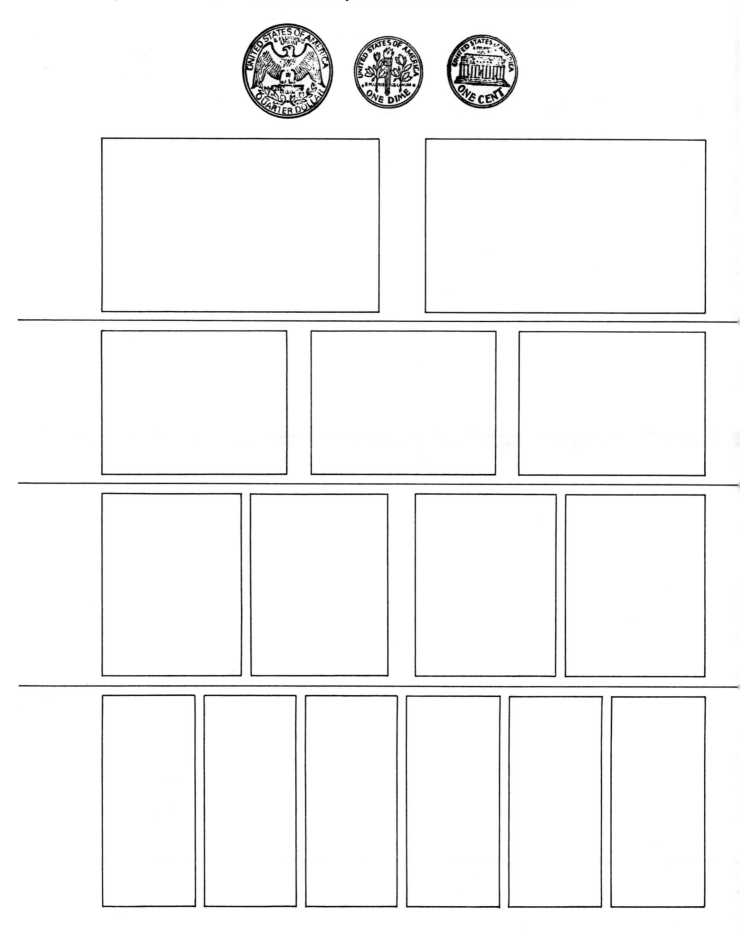

Sarah receives money from her mom for doing chores around the house. To buy different things, she must divide up her savings. Help her by stamping an equal value in each space.

Rene's mom gives Rene an amount of money. Then Rene rolls a die. She must divide the value of the coins equally by the number that appears on the die. If she does the problem correctly, her mom lets Rene keep a portion. Help Rene solve the problems below.

Rene rolls a 5

Rene rolls a 4

What number do you think Rene would like to roll? _____

John and Sarah start a lemonade stand. Help them calculate how much money they will have on Friday. Fill in the chart by stamping the correct amounts in the blank boxes. (Remember each day they start with the money total from the day before!)

	Start day with	Take in	End day with
Monday (Sunny)			$ ____
Tuesday (rainy)			$ ____
Wednesday (hot)			$ ____
Thursday (partly cloudy)			$ ____
Friday (cloudy)			$ ____

Stamp exactly 3 coins to total the values listed.

$0.16

$0.45

$0.80

$0.40

$0.76

$0.52

Stamp exactly 3 coins to total the values listed.

$0.80

$0.65

$1.00

$1.50

$1.25

$1.10

Stamp exactly 6 coins to total the values listed.

$0.51

$0.28

$0.67

$0.72

$0.65

$1.00

Stamp exactly 6 coins to total the values listed.

$1.26

$0.41

$2.60

$0.95

$1.06

$2.50

Stamp 7 coins to make the amount listed.

$0.93

$2.10

$0.37

$1.60

$0.86

$1.27

Stamp 8 coins to make the amount listed.

$.91

$1.30

$.38

$1.85

$.84

$2.05

Write the letter that tells the correct change in the space provided.

	Had	Spent	Your change would be
1. _____	$1.00	$0.89	a)
2. _____	$1.00	$0.23	b)
3. _____	$1.00	$.57	c)
4. _____	$2.00	$1.32	d)
5. _____	$2.00	$1.45	e)
6. _____	$2.00	$1.68	f)

Stamp the correct change you would give a customer if you were the cashier in the following situations.

Customer gives $1.00 Your purchase $0.79
Customer gives $1.00 Your purchase $0.07
Customer gives $1.00 Your purchase $0.81
Customer gives $1.00 Your purchase $0.12
Customer gives $1.00 Your purchase $0.62
Customer gives $1.00 Your purchase $0.35

Stamp the correct change you would give a customer if you were the cashier in the following situations.

Customer gives $2.00
Your purchase $1.47

Customer gives $3.50
Your purchase $2.36

Customer gives $3.50
Your purchase $2.43

Customer gives $5.00
Your purchase $3.88

Customer gives $5.00
Your purchase $4.01

Customer gives $6.00
Your purchase $5.25

Sometimes people give cashiers pennies or other change in order to avoid getting a large number of coins. Notice this as you stamp the change in this exercise.

Amount of Purchase $0.27
Give Cashier $1.02

Amount of Purchase $0.31
Give Cashier $1.01

Amount of Purchase $0.76
Give Cashier $1.01

Amount of Purchase $3.29
Give Cashier $4.04

Amount of Purchase $2.92
Give Cashier $3.02

Amount of Purchase $3.28
Give Cashier $5.28

Stamp the change to be given.
Keep in mind that people want as few coins as possible!

Amount of Purchase $1.16
Give Cashier $2.06

Amount of Purchase $1.47
Give Cashier $2.50

Amount of Purchase $3.36
Give Cashier $3.51

Amount of Purchase $3.27
Give Cashier $4.02

Amount of Purchase $0.86
Give Cashier $1.06

Amount of Purchase $4.02
Give Cashier $5.10

Did the cashier give the correct change? Circle YES or NO

Amount of Purchase $2.16
Give Cashier $2.50

YES NO

Amount of Purchase $1.89
Give Cashier $2.00

YES NO

Amount of Purchase $3.23
Give Cashier $4.03

YES NO

Amount of Purchase $5.44
Give Cashier $6.00

YES NO

Amount of Purchase $7.36
Give Cashier $8.06

YES NO

Amount of Purchase $9.17
Give Cashier $10.03

YES NO

It is always important to check to make sure you have the right change before you leave a store. Check the change below and indicate by circling YES or NO

Amount of Purchase $0.29
Give Cashier $1.00

YES   NO

Amount of Purchase $0.43
Give Cashier $0.50

YES   NO

Amount of Purchase $0.66
Give Cashier $1.00

YES   NO

Amount of Purchase $0.26
Give Cashier $1.01

YES   NO

Amount of Purchase $4.71
Give Cashier $5.00

YES   NO

Amount of Purchase $6.12
Give Cashier $7.02

YES   NO

Did the cashier give the correct change? Circle YES or NO

Amount of Purchase $3.36
Give Cashier $4.01

YES NO

Amount of Purchase $0.84
Give Cashier $1.00

YES NO

Amount of Purchase $4.26
Give Cashier $5.00

YES NO

Amount of Purchase $1.41
Give Cashier $2.01

YES NO

Amount of Purchase $5.45
Give Cashier $6.00

YES NO

Amount of Purchase $4.39
Give Cashier $5.00

YES NO

Cashiers "count out" change for their customers. Look at the example and "count out" change for the remaining situations. Stamp the coins and write the tally below.

Amount of Purchase $8.59 Gave Cashier $9.00	$8.60;	$8.70;	$8.75;	$9.00

Amount of Purchase $3.47
Gave Cashier $4.00

___ ___ ___ ___ ___ ___

Amount of Purchase $2.91
Gave Cashier $3.00

___ ___ ___ ___ ___ ___

Amount of Purchase $0.35
Give Cashier $1.00

___ ___ ___ ___ ___

Amount of Purchase $4.78
Give Cashier $5.00

___ ___ ___ ___ ___ ___

Amount of Purchase $5.65
Give Cashier $6.00

___ ___ ___ ___ ___ ___

Perform the operations using the value of the coins.

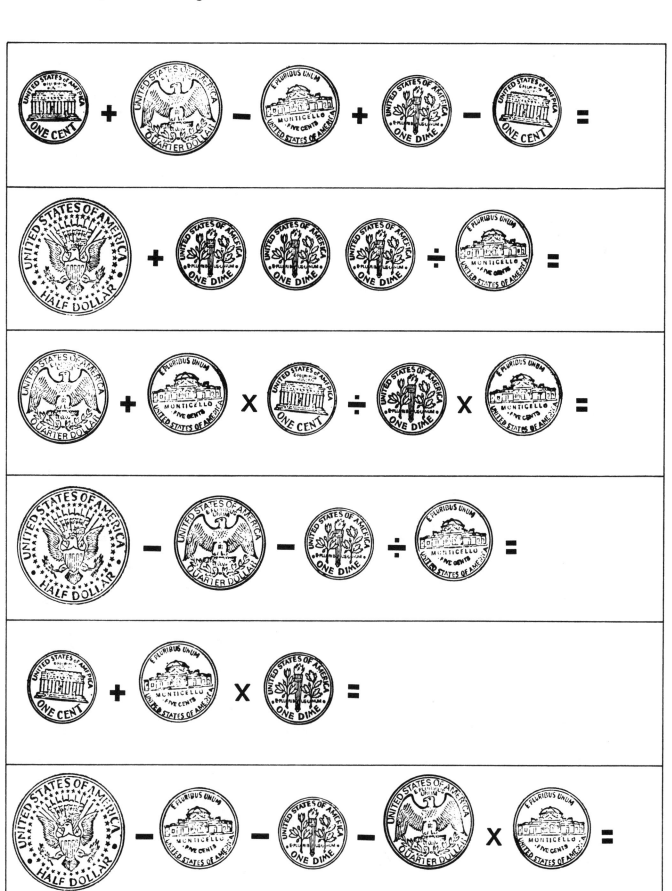

Perform the operations using the value of the coins.

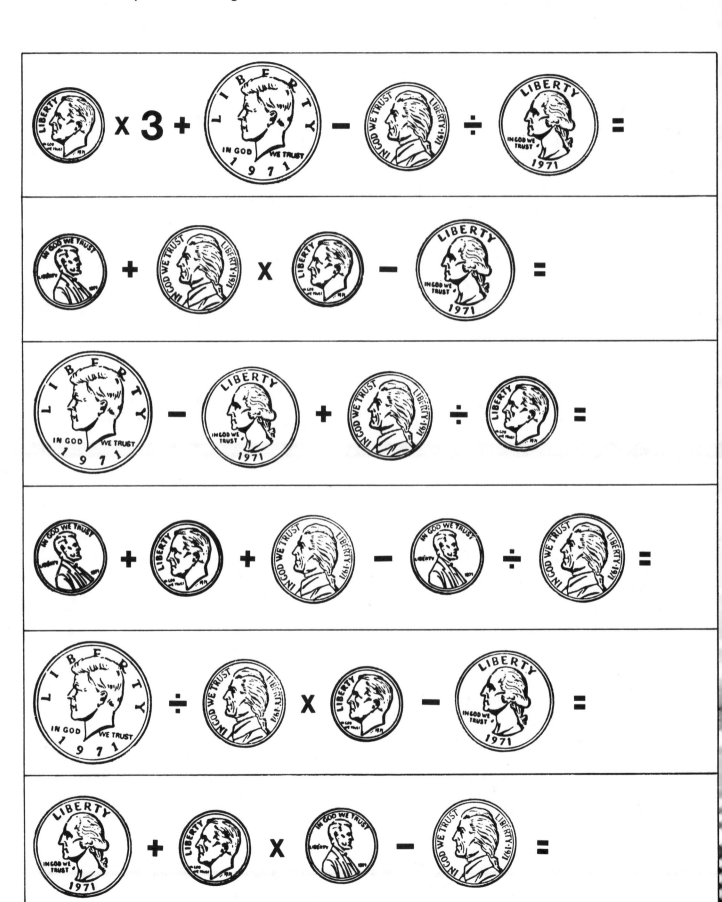

Perform the operations using the value of the coins.

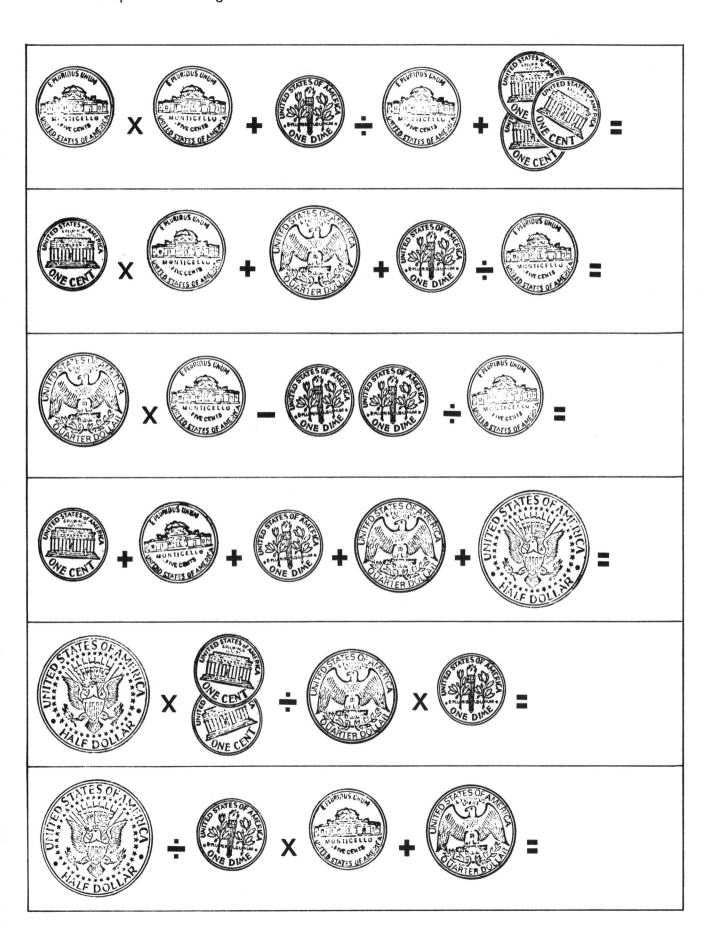

Jim Clerk had 6 price tags he had to label. However, he lost his list. What he did know was:

A. All prices were multiples of 5 and could only be made by stamping nickel or dimes in them.
B. Tag 1 has the smallest price
C. No two prices are the same, nor is any price greater than 50¢ or less than 10¢.
D. The value of price tag 6 is half the value of tag 3.
E. Tag 4 has nine coins but its value is less than tag 3.
F. The value of tag 2 is three times the value of tag 1.
G. The value of tag 5 is half the value of tag 2.

Now can you help Jim Clerk stamp the correct prices on the tags.

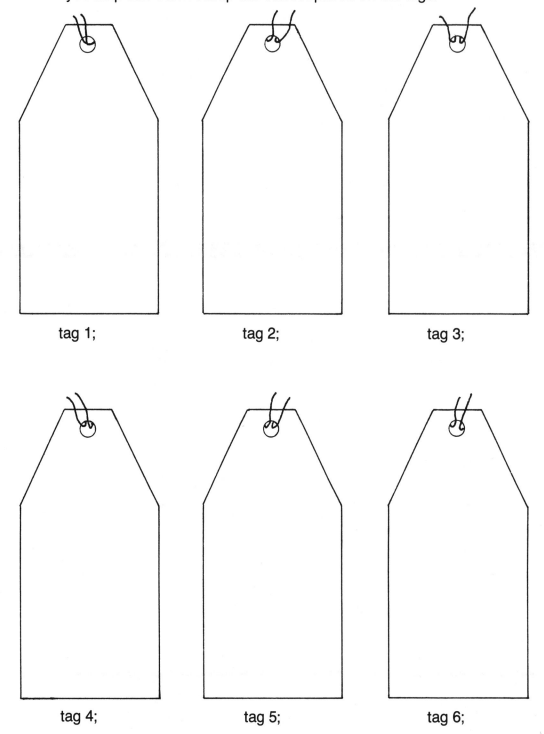

tag 1;   tag 2;   tag 3;

tag 4;   tag 5;   tag 6;

Spacey Sally has been given 2 quarters, 10 dimes, 5 nickels and 8 pennies by Mischief-maker Mike. With them was a list of clues. Help Sally by reading the clues and stamping the coins in the correct circle.

1. Pennies go in circles 3 and 6; Quarters in circles 1 and 2.
2. Nickels go in circles 2, 4, and 5—Circle 4 only contains nickels.
3. Dimes go in circles 1, 2, 3 and 5. Circles 1, 2, 3 all contain the same number of dimes—Circle 5 contains double that number.
4. Circles 1 and 4 contain the same number of coins which is one less than circles 2 and 3 which both contain four coins.
5. No circle has less than three nor more than six coins.
6. No circle has less than 6¢ nor more than 50¢.
7. Circles 1 and 5 have the same value. However, circle 1 has fewer coins in it.

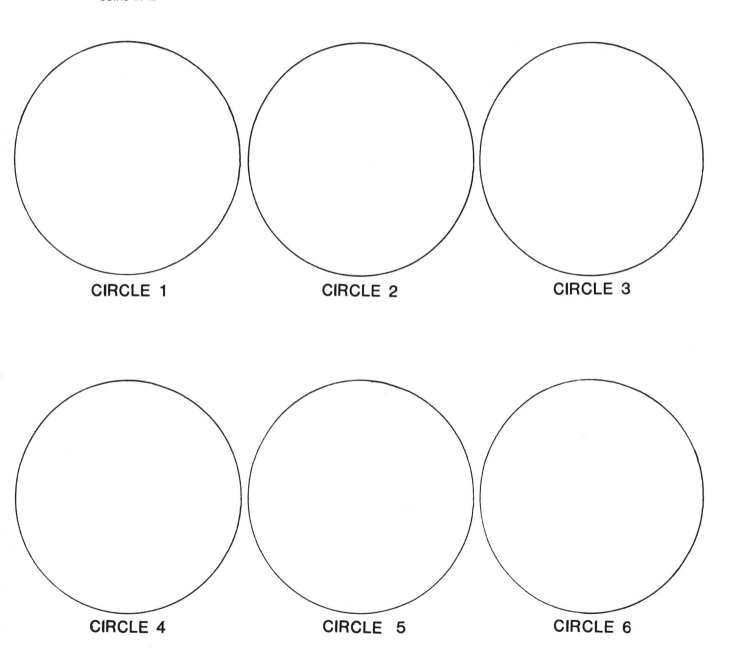

CIRCLE 1 CIRCLE 2 CIRCLE 3

CIRCLE 4 CIRCLE 5 CIRCLE 6

ANSWER KEY

Answers for Page 45

end;	55¢;
55¢;	60¢;
60¢;	$1.10;
$1.10;	$1.40;
$1.40;	$1.48

Answers for Page 64

tag 1 = 10¢
tag 2 = 30¢
tag 3 = 50¢
tag 4 = 45¢
tag 5 = 15¢
tag 6 = 25¢

Answers for Page 65

circle 1 = 45¢ 1Q; 2D
circle 2 = 50¢ 1Q; 2D; 1N
circle 3 = 22¢ 2D; 2P
circle 4 = 15¢ 3N
circle 5 = 45¢ 4D; 1N
circle 6 = 6¢ 6P

Answers for Page 3

1. dime
2. In God We Trust—Liberty
3. nickel
4. 1¢ or $0.01
5. Washington
6. Quarter & Half-dollar
7. Lincoln Memorial